WEIRD ANIMAL SCIENCE

STINGING ANIMALS

THE BREAKDOWN

VIRGINIA LOH-HAGAN

45TH PARALLEL PRESS

Published in the United States of America by Cherry Lake Publishing Group
Ann Arbor, Michigan
www.cherrylakepublishing.com

Photo Credits: © Mirko Graul/Shutterstock, cover, 1; © bsd studio/Shutterstock, cover, 1, 8; © EcoPic/iStock, 6; © Ajay Shrivastava/Shutterstock, 7; © Frank Reiser/Shutterstock, 9; © Irina Kozorog/Shutterstock, 10; © Gewoldi/iStock, 13; © Tim the Finn/Shutterstock, 14; © Katherine OBrieni/iStock, 15; © Yann hubert/Shutterstock, 16; © Grilleau Nicolas/iStock, 18; © plains-wanderer/Shutterstock, 21

Graphic Element Credits: Graphic Element Credits: Cover, multiple interior pages: © WindAwake/Shutterstock; © Theo_hrm/Shutterstock; © Rodin Anton/Shutterstock; © Petr Vaclavek/Shutterstock; © Nikolaeva/Shutterstock

45th Parallel Press is an imprint of Cherry Lake Publishing Group.

Library of Congress Cataloging-in-Publication Data has been filed and is available at catalog.loc.gov

Cherry Lake Publishing Group would like to acknowledge the work of the Partnership for 21st Century Learning, a Network of Battelle for Kids. Please visite Battelle for Kids online for more information.

Printed in the United States of America

Note from publisher: Websites change regularly, and their future contents are outside of our control. Supervise children when conducting any recommended online searches for extended learning opportunities.

Dr. Virginia Loh-Hagan is an author, university professor, and former classroom teacher. She's currently the Director of the Asian Pacific Islander Desi American Resource Center at San Diego State University. She loves science-fiction stories. She lives in San Diego with her very tall husband and very naughty dogs.

TABLE OF CONTENTS

BREAK IT DOWN

1. DECODE IT

- Look at each letter.
- Look at letter patterns.
- Blend sounds.
- Try other possible sounds.

2. BUILD VOCABULARY

- Listen to the word.
- Match the sounds to a word you know.
- Look for context clues for words you don't know.
- Look up the meaning in a glossary or dictionary.

3. MAKE CONNECTIONS

- Name other examples.
- Draw the idea.
- Compare ideas.
- Ask questions.

TEST IT OUT

WORDS YOU'LL LEARN

Look over the list of vocabulary words. Are there any you already know?

barbs
glands
stingers
strike
triggers
venom

GET STARTED

Letter sounds can change when they blend together in words. Pay attention to how consonants blend in words.

- Look at the letters.
- Look for 2 consonants right next to each other.
- Try blending the sounds.

The word *stingers* starts with the consonants *s* and *t*. The *s* and *t* each still make a sound. The sound is just blended together. *Stingers* is pronounced: STING-uhrz.

- The *s* says "sss."
- The *t* says "ttt."
- Together, they say "st."

Animals sting to protect themselves.

Stingers are sharp. They look like needles.

Bees have stingers.

They use them to pierce skin.

They release a **venom**.

Venom is poison that's injected.

Venom travels to the bloodstream.

It **triggers** pain nerves. This means sets off.

This causes a burning feeling.

It causes redness and swelling.

DECODE IT

The word *triggers* starts with the consonants *t* and *r*.

- The *t* says "ttt." The r says "rrr." Together, they say "tr."
- The 1st vowel is a short *i*.
- The next vowel is an *r*-controlled *e*. It says "uhr."
- *Triggers* is pronounced TRIH-guhrz.

A bee's venom comes from **glands**.

BUILD VOCABULARY

Glands (GLANDZ) are organs that produce substances like venom.

- Animals have lots of glands.
- Humans have more than 40 glands in their bodies!
- Other forms: gland, glandular, glandless

Glands are organs that produce substances such as venom.

Stingrays are flat fish.

They swim at the bottom of warm waters.

They bury themselves in sand.

They hide. They're hard to see.

They attack anyone who steps on them.

They whip their tails around.

Their tails have stingers.

The stingers **strike** and dig in deep. Strike means to pierce.

DECODE IT

The word *strike* starts with 3 consonants: *s*, *t*, and *r*.

- The *s* says "sss." The *t* says "ttt." The *r* says "rrr." Together, they say "str."
- The vowel sound is a long *i*.
- *Strike* is pronounced STRIYK.

They inject venom.

Their venom is located underneath their stingers.

Some stingers can be 19 inches (48 centimeters) long!

Stingers can be different. Some have spurs. Some have **barbs**.

MAKE CONNECTIONS

Some animals have barbs to protect themselves.

- Some plants have barbs. They stick to fur and clothes. How might this help plants?
- Humans created barbed wire. How is it used like animals use their barbs?

Barbs are spikes that hook into things. Barbs hurt more when pulled out.

PUT IT TOGETHER

Detail:
Stingrays can sting and release a venom.

Detail:
Bees can sting and release a venom.

Detail:
Some animals sting in different ways to protect themselves.

Main Idea:
Animals sting to protect themselves.

Detail:
Some stingers have spurs or barbs.

Detail:
Stingers are sharp and look like needles.

LEARN MORE

BOOKS

Jenkins, Steve. *The Animals Book*. Boston, MA: HMH Books for Young Readers, 2013.

Lerwill, Ben, and Sarah Walsh (illus.). *WildLives: 50 Extraordinary Animals That Made History*. New York, NY: Atheneum Books for Young Readers, 2020.

Watson, Galadriel, and Samantha Dixon (illus.) *Running Wild: Awesome Animals in Motion*. Toronto, ON: Annick Press, 2020.

ONLINE

Search these online sources with an adult:

- National Geographic Kids | Animals
- Britannica Kids | Animal Kingdom

GLOSSARY

barbs (BARBS) spikes used to hook into things

glands (GLANDZ) body organs that secrete substances

stingers (STING-uhrz) sharp organs that pierce and share poison

strike (STRIYK) to pierce

triggers (TRIH-guhrz) sets off

venom (VEH-nuhm) poison that is injected under the skin

INDEX